I0503724

Boiler Operator Handbook

A Comprehensive Guide To Boiler Operations And Maintenance

Copyright@2023

Cary Leighton

Table Of Content

Chapter 1: Introduction

A. Importance Of Boilers In Industrial Processes

Boilers play a crucial role in various industrial processes and are essential for the efficient operation of many industries. Here are some key reasons why boilers are important in industrial processes:

1. Steam Generation: Boilers are primarily used for steam generation, which is a vital component in numerous industrial processes. Steam is used for heating, power generation, sterilization, humidification, and various other applications across industries.

2. Heat Transfer: Boilers facilitate the transfer of heat from fuel combustion to water or other fluids, which results in the production of steam or hot water. This heat transfer is essential for heating processes, such as space heating in buildings, heating in manufacturing processes, and maintaining specific temperatures in industrial equipment.

3. Power Generation: Boilers are extensively utilized in power generation plants, where they produce steam to drive steam turbines and generate electricity. Power plants, including coal-fired, gas-fired, and nuclear power plants, heavily rely on boilers to convert thermal energy into mechanical energy and eventually electrical energy.

4. Industrial Processes: Many industrial processes require heat for manufacturing, processing, or chemical reactions. Boilers provide the necessary heat to initiate and sustain these processes. Industries such as food processing, pharmaceuticals, chemical production, oil refineries, textiles, and paper mills heavily depend on boilers to meet their specific heating and processing needs.

5. Heating and HVAC Systems: Boilers are commonly used for heating applications in buildings, including commercial and residential spaces. They provide centralized heating by distributing hot water or steam through pipes and radiators, ensuring comfortable indoor temperatures during colder months.

6. Hot Water Supply: Boilers are instrumental in providing hot water for various purposes, such as bathing, cleaning, and sanitation in industrial facilities, hotels, hospitals, and other institutions. They ensure a constant supply of hot water at the desired temperature, catering to the specific needs of each application.

7. Process Control and Automation:
Modern boilers are equipped with advanced control systems and automation technology, allowing precise regulation of temperature, pressure, and fuel combustion. This enables efficient and safe operation, ensuring optimal performance and minimizing energy waste.

8. Environmental Considerations: Boilers also play a role in environmental sustainability. Advancements in boiler technology and the adoption of cleaner fuel options, such as natural gas and biomass, contribute to reduced emissions and improved air quality. Additionally, waste heat recovery systems associated with boilers can help recover and utilize waste heat, enhancing energy efficiency and reducing greenhouse gas emissions.

B. Role And Responsibilities Of A Boiler Operator

The role of a boiler operator is critical in ensuring the safe and efficient operation of boilers in industrial settings. Boiler operators are responsible for overseeing the operation, maintenance, and troubleshooting of boilers to ensure proper functioning and adherence to safety regulations. Here are the key roles and responsibilities of a boiler operator:

1. Boiler Operation:

- Start up and shut down boilers following specified procedures.

- Monitor and control boiler operation parameters such as temperature, pressure, fuel and air flow rates, water levels, and steam production.

- Adjust boiler controls and settings to maintain optimal boiler performance.

- Monitor gauges, meters, and computerized controls to identify abnormalities or malfunctions in the boiler system.

- Ensure the efficient and safe combustion of fuels, such as natural gas, coal, oil, or biomass, to generate steam or hot water.

2. Safety and Compliance:

- Follow strict safety protocols and procedures to ensure a safe working environment.

- Monitor and maintain appropriate water levels, pressure, and temperature to prevent overheating or boiler explosions.

- Conduct regular inspections to identify and rectify potential safety hazards.

- Ensure compliance with relevant codes, standards, and regulations, such as those related to boiler operation, emissions, and environmental protection.

- Perform routine maintenance tasks, including cleaning, lubricating, and inspecting boiler equipment, to prevent equipment failures and ensure safe operations.

3. Maintenance and Troubleshooting:

- Perform routine maintenance activities, such as cleaning, repairing, and replacing boiler components and equipment.

- Troubleshoot and diagnose boiler malfunctions, identify the root causes, and implement corrective actions.

- Conduct regular inspections of boiler systems to identify leaks, corrosion, or other issues that may impact performance or safety.

- Coordinate with maintenance staff or contractors for major repairs or equipment upgrades.

- Keep accurate records of maintenance activities, inspections, and repairs.

4. Communication and Documentation:

- Communicate effectively with other team members, supervisors, and maintenance personnel regarding boiler operations and maintenance activities.

- Maintain accurate logs and records of boiler operations, including fuel consumption, water levels, and steam production.

- Report any accidents, incidents, or unusual events to the appropriate authorities.

- Document and report any deviations from standard operating procedures or safety protocols.

5. Training and Continuous Learning:

- Stay updated with the latest industry regulations, safety practices, and technological advancements in boiler operations.

- Attend training sessions, workshops, or seminars to enhance knowledge and skills related to boiler operations.

- Provide guidance and training to other boiler operators or new employees regarding safe and efficient boiler operations.

Chapter 2: Boiler Fundamentals

A. Overview Of Boilers And Their Components

Boilers are vessels designed to generate steam or hot water by heating a fluid, typically water, to high temperatures using various fuel sources. They are a critical component in many industrial processes and heating systems. Here is an overview of boilers and their key components:

1. Boiler Shell: The boiler shell is a cylindrical or rectangular pressure vessel that contains the water or steam. It provides a housing for the combustion chamber and other components.

2. Combustion Chamber: The combustion chamber is where fuel, such as natural gas, oil, coal, or biomass, is burned to generate heat. It is designed to facilitate efficient combustion and transfer the released heat to the surrounding water or fluid.

3. Burners: Burners are responsible for the controlled combustion of fuel. They mix the fuel and air in the right proportions and ignite the mixture in the combustion chamber. Burners can be of various types, including atmospheric burners, power burners, or low-NOx burners, depending on the specific application and environmental requirements.

4. Heat Exchanger: The heat exchanger is a key component that transfers heat from the combustion gases to the water or fluid. It consists of a series of tubes or passages where the hot gases flow, allowing for heat transfer through conduction and convection.

5. Water or Steam Circulation System:
Boilers have a system for circulating water or steam within the vessel. In water-tube boilers, water circulates through tubes located inside the boiler, while in fire-tube boilers, the hot gases pass through tubes immersed in the water. This circulation system facilitates the transfer of heat and the generation of steam or hot water.

6. Control Systems: Boilers are equipped with control systems that regulate and monitor various parameters, including temperature, pressure, fuel and air flow rates, water level, and safety devices. These systems ensure safe and efficient boiler operation by maintaining optimal conditions and responding to changes or anomalies.

7. Safety Devices: Boilers incorporate safety devices to prevent overpressure, overheating, and other potentially hazardous conditions. These devices include pressure relief valves, temperature and pressure gauges, low water cutoff switches, flame safeguards, and fuel safety shutoff valves. They help maintain the integrity and safe operation of the boiler system.

8. Pumps and Valves: Pumps and valves are essential for the proper circulation and control of water, steam, and other fluids within the boiler system. They ensure adequate flow rates, pressure control, and the isolation of different sections of the system for maintenance or repairs.

9. Insulation: Boilers are typically insulated to minimize heat loss and improve energy efficiency. Insulation materials such as refractory materials, ceramic fiber, or insulation jackets help retain heat within the boiler, reducing energy consumption and improving safety.

10. Chimney or Flue: The chimney or flue is the outlet for the combustion gases and exhaust. It provides a pathway for the gases to be safely vented out of the boiler and released into the atmosphere.

It's important to note that boiler designs and components may vary based on the type and purpose of the boiler, such as fire-tube boilers, water-tube boilers, package boilers, or industrial boilers. Additionally, specific applications may require additional components or systems, such as economizers, air preheaters, deaerators, or water treatment systems, to optimize performance and meet specific operational needs.

B. Types Of Boilers And Their Applications

There are various types of boilers available, each with its own design, characteristics, and applications. The choice of boiler type depends on factors such as the intended purpose, fuel availability, efficiency requirements, and industry-specific considerations. Here are some common types of boilers and their applications:

1. Fire-Tube Boilers:

  - Application: Fire-tube boilers are widely used in small to medium-scale industrial processes, heating applications, and commercial buildings.

  - Description: In fire-tube boilers, hot gases produced by the combustion of fuel flow through tubes immersed in water. The transfer of heat to the water results in the production of steam or hot water.

2. Water-Tube Boilers:

- Application: Water-tube boilers are commonly used in large-scale industrial applications, power generation plants, and high-pressure steam processes.

- Description: Water-tube boilers have water-filled tubes that are externally heated by hot combustion gases. The heated water converts into steam within the tubes.

3. Package Boilers:

- Application: Package boilers are compact, pre-engineered boiler units commonly used in various industries, including pharmaceuticals, chemical plants, food processing, and hospitals.

- Description: Package boilers are factory-assembled units with all necessary components, including the boiler, control systems, pumps, and valves, housed in a single package. They are designed for quick installation and ease of operation.

4. Industrial Boilers:

- Application: Industrial boilers are large-scale boilers used in heavy industries such as oil refineries, petrochemical plants, paper mills, textile factories, and power generation plants.

- Description: Industrial boilers are designed to handle high pressures, temperatures, and large steam capacities to meet the demands of industrial processes, power generation, or heating applications.

5. Steam Boilers:

- Application: Steam boilers are widely used in power plants, manufacturing processes, hospitals, district heating systems, and HVAC applications.

- Description: Steam boilers produce steam that can be used for heating, power generation, mechanical processes, or sterilization. They come in various configurations and sizes to meet different application requirements.

6. Hot Water Boilers:

- Application: Hot water boilers are commonly used in commercial buildings, hotels, schools, and residential heating systems.

- Description: Hot water boilers heat water to supply hot water for space heating or domestic use. They are typically used in systems that require lower temperatures compared to steam boilers.

7. Biomass Boilers:

- Application: Biomass boilers are used in industries where biomass resources, such as wood pellets, agricultural residues, or dedicated energy crops, are available.

- Description: Biomass boilers burn biomass fuels to produce heat or steam. They are considered a renewable and sustainable alternative to traditional fossil fuel boilers, promoting environmental sustainability.

8. Electric Boilers:

- Application: Electric boilers are commonly used in residential buildings, small-scale industrial processes, and heating systems where electricity is the primary energy source.

- Description: Electric boilers use electricity to heat water or generate steam. They are efficient, compact, and do not produce emissions, making them suitable for specific applications and areas with strict environmental regulations.

C. Boiler Construction And Design Principles

Boiler construction and design principles play a crucial role in ensuring the safe and efficient operation of boilers. Here are the key considerations and principles involved in the construction and design of boilers:

1. Pressure Vessel Design: Boilers are pressure vessels subject to high internal pressure. Their construction must adhere to applicable codes and standards, such as the ASME Boiler and Pressure Vessel Code. The design considers factors like maximum allowable working pressure, material strength, corrosion resistance, and structural integrity.

2. Boiler Shell: The boiler shell is the main body of the boiler and is typically constructed using steel or alloy materials. It provides a robust enclosure for the combustion chamber, tubes, and other components. The shell must be adequately reinforced to withstand internal pressure and external forces.

3. Combustion Chamber Design: The combustion chamber design aims to facilitate efficient fuel combustion while ensuring safe and controlled heat transfer. Factors like residence time, turbulence, fuel-air mixing, and temperature distribution are considered to achieve optimal combustion efficiency and minimize emissions.

4. Tube Design: For water-tube boilers, the design of tubes is crucial for efficient heat transfer and structural integrity. Tubes may be straight or bent, and their arrangement should allow for effective circulation and thermal expansion. Proper sizing, material selection, and tube spacing are important considerations.

5. Heat Exchanger Surface Area: The heat exchanger surface area determines the heat transfer capacity of the boiler. Sufficient surface area is required to transfer heat from the combustion gases to the water or steam effectively. The surface area is influenced by factors like boiler capacity, operating conditions, and desired efficiency.

6. Fuel and Air Delivery System: The design of the fuel and air delivery system ensures proper combustion. It involves considerations such as fuel and air flow rates, fuel atomization or pulverization, combustion air supply, and control mechanisms to maintain the desired fuel-air ratio for efficient combustion.

7. Water or Steam Circulation System:
The design of the water or steam circulation system in boilers varies based on the type and purpose of the boiler. It involves considerations like flow rates, pressure drops, pipe sizing, and the arrangement of tubes or passages to promote effective heat transfer and minimize pressure losses.

8. Insulation and Refractory Materials:
Boiler construction incorporates insulation materials to minimize heat loss and improve energy efficiency. Refractory materials are used to line the combustion chamber and other high-temperature areas to protect the boiler structure from excessive heat and ensure durability.

9. Safety Considerations: Boiler design includes the incorporation of safety features and devices to prevent accidents, such as pressure relief valves, temperature and pressure gauges, low water cutoff switches, and flame safeguards. These safety devices ensure the protection of the boiler and personnel.

10. Control Systems: Boiler design includes provisions for control systems to monitor and regulate various parameters like temperature, pressure, fuel and air flow rates, and water level. Advanced control systems may include automation, sensors, and safety interlocks to ensure safe and efficient boiler operation.

11. Environmental Considerations: Boiler design principles increasingly focus on environmental sustainability. Efforts are made to reduce emissions through optimized combustion, flue gas treatment systems, and the incorporation of pollution control technologies.

Boiler construction and design principles are continuously evolving to meet industry standards, improve efficiency, enhance safety, and reduce environmental impact. Manufacturers, engineers, and regulatory bodies collaborate to ensure that boiler designs are robust, reliable, and capable of meeting the specific operational needs of different industries and applications.

D. Boiler Efficiency And Heat Transfer

Boiler efficiency and heat transfer are closely related and crucial factors in optimizing the performance and energy efficiency of boilers. Let's explore how heat transfer affects boiler efficiency and the various mechanisms involved:

1. Heat Transfer in Boilers:

Heat transfer in boilers occurs through three primary mechanisms:

a. Conduction: Conduction is the transfer of heat through direct contact between solids. In boilers, conduction takes place between the combustion gases and the boiler tubes or heat exchanger surfaces, allowing heat to pass from the hot gases to the water or steam.

b. Convection: Convection involves the transfer of heat through the fluid's motion. In boilers, convection occurs as hot gases rise and cooler fluids, such as water or steam, take their place. This convective heat transfer enhances the overall heat transfer efficiency.

c. Radiation: Radiation is the process by which heat is transferred through the emission and propagation of electromagnetic waves. In boilers, radiation occurs as heat is emitted from the hot combustion gases and is absorbed by the surrounding surfaces, including the boiler walls, tubes, and other components.

2. Boiler Efficiency:

Boiler efficiency is a measure of how effectively a boiler converts fuel into usable heat energy. It is typically expressed as a percentage and is influenced by various factors:

a. Combustion Efficiency: Combustion efficiency is a measure of how efficiently fuel is burned in the boiler's combustion chamber. It depends on factors like fuel-air mixing, combustion chamber design, and the combustion process control. Higher combustion efficiency leads to reduced fuel waste and increased boiler efficiency.

b. Thermal Efficiency: Thermal efficiency is a measure of how effectively a boiler converts fuel's energy content into usable heat energy. It takes into account both the combustion efficiency and the heat transfer efficiency of the boiler. Improving thermal efficiency involves maximizing heat transfer and minimizing energy losses through proper insulation, optimized combustion, and efficient heat exchanger design.

c. Overall Efficiency: Overall efficiency considers all energy losses, including those from radiation, flue gas losses, blowdown losses, and other operational losses. Maximizing overall efficiency involves minimizing these losses through proper insulation, efficient combustion, and effective control systems.

3. Factors Affecting Heat Transfer and Efficiency:

Several factors influence heat transfer and overall boiler efficiency:

a. Surface Area: Increasing the heat transfer surface area within the boiler enhances heat transfer efficiency. This can be achieved through the design and arrangement of tubes, heat exchanger surfaces, and finned surfaces.

b. Temperature Difference: The greater the temperature difference between the hot gases and the water or steam, the higher the heat transfer rate. Optimizing temperature differences by controlling combustion and fluid temperatures can improve efficiency.

c. Clean Heat Transfer Surfaces: Deposits, scaling, or fouling on heat transfer surfaces reduce heat transfer efficiency. Regular cleaning and maintenance of boiler surfaces help maintain optimal heat transfer rates and prevent efficiency losses.

d. Fluid Flow: Proper fluid flow rates and velocity are important for efficient heat transfer. Adequate flow ensures that heat is effectively carried away from the heat source to the heat sink.

e. Combustion Control: Efficient combustion control ensures that the right amount of fuel and air is supplied for optimal combustion. Proper control of excess air and maintaining the stoichiometric fuel-air ratio improves combustion efficiency and overall boiler efficiency.

f. Boiler Sizing and Matching: Properly sizing the boiler to the heat load requirements ensures efficient operation. Oversized or undersized boilers can lead to reduced efficiency and increased energy consumption.

Optimizing heat transfer and boiler efficiency involves a combination of design considerations, operational practices, and maintenance procedures. By maximizing heat transfer rates and minimizing energy losses, boilers can operate more efficiently, resulting in reduced fuel consumption and lower operating costs. Regular monitoring, performance evaluation, and maintenance are essential for sustained

Chapter 3: Boiler Operation And Safety

A. Boiler Start-Up And Shutdown Procedures

Boiler start-up and shutdown procedures are essential for safe and efficient operation, as they ensure proper system readiness and minimize the risk of equipment damage. Here are the general steps involved in boiler start-up and shutdown:

Boiler Start-Up Procedure:

1. Pre-Start Checks:

- Ensure that the boiler area is clear of any obstructions or hazards.

- Check the boiler water level, ensuring it is at the correct level for operation.

- Verify that all valves, switches, and control devices are in the proper position.

- Ensure that the fuel supply is available and the fuel valves are open.

- Inspect the boiler and surrounding area for any leaks, damages, or abnormal conditions.

2. Boiler Purge:

- Open the main fuel valve and ensure proper fuel flow to the burner.

- Purge the boiler combustion chamber by opening the air vent valves to release any air or gases.

- Close the vent valves once a steady flow of fuel is established.

- Follow the specific manufacturer's instructions for purging and lighting the burner.

3. Ignition and Start-Up:

- Follow the manufacturer's instructions for igniting the burner flame.

- Monitor the flame and ensure it is stable and properly established.

- Gradually increase the boiler pressure and temperature, following the recommended ramp-up rates.

4. System Checks:

- Monitor and adjust the combustion process, ensuring optimal fuel-air mixture and combustion efficiency.

- Check and adjust the water level, pressure, and temperature according to system requirements.

- Verify that all safety devices, such as pressure relief valves and low water cutoff switches, are functioning correctly.

- Monitor other parameters, such as flue gas temperature, stack emissions, and operating pressures, as required.

Boiler Shutdown Procedure:

1. Reduce Load:
  - Gradually reduce the boiler load by adjusting the fuel and air supply to match the reduced demand.
  - Monitor and record operating parameters during the load reduction process.

2. Blowdown:
  - Perform a bottom blowdown to remove sediment and impurities from the boiler.
  - Follow proper procedures to safely drain a portion of the boiler water.
  - Conduct surface blowdown if necessary to remove suspended solids.

3. Cooling:
  - Allow the boiler to cool down gradually to prevent thermal stresses.

- Maintain the water circulation and monitor the water level during the cooling process.

- Open the boiler vent valves to allow for the release of trapped air or gases.

4. Shutdown:

- Shut off the fuel supply to the burner and extinguish the flame according to manufacturer instructions.

- Close all valves, switches, and control devices related to the boiler operation.

- Conduct a final inspection of the boiler and surrounding area for any abnormalities or hazards.

B. Boiler Fuel And Combustion Systems

Boiler fuel and combustion systems play a critical role in the efficient and reliable operation of boilers. The selection of the appropriate fuel and the design of the combustion system have a significant impact on boiler performance, emissions, and overall efficiency. Here are some key aspects related to boiler fuel and combustion systems:

1. Fuel Types:

- Solid Fuels: Solid fuels such as coal, wood, biomass, and solid waste can be used as boiler fuels. They are typically in solid or pelletized form and require specific handling and feeding systems.

- Liquid Fuels: Liquid fuels include petroleum-based fuels such as diesel, fuel oil, and heavy fuel oil. They are commonly used in applications where natural gas or other gaseous fuels are not available or economical.

- Gaseous Fuels: Gaseous fuels, such as natural gas, propane, and biogas, are increasingly popular due to their clean-burning properties and ease of combustion.

2. Fuel Handling and Storage:

- The fuel handling systems differ based on the specific type of fuel utilized. Solid fuels require storage areas, fuel handling equipment (e.g., conveyors, feeders), and proper handling procedures to ensure a consistent and reliable fuel supply.

- Liquid fuels are typically stored in tanks and require appropriate piping and pump systems for delivery to the boiler.

- Gaseous fuels are supplied through pipelines, and the gas pressure may require regulation before entering the combustion system.

3. Combustion Air Supply:

- Combustion requires the proper supply of air or oxygen for efficient fuel combustion. Air supply systems, such as fans or blowers, provide the necessary air to the combustion process.

- The air-to-fuel ratio needs to be carefully controlled to achieve complete combustion while minimizing excess air, which can result in energy losses.

4. Combustion Process:

- The combustion process involves the rapid oxidation of fuel in the presence of oxygen. The combustion chamber or burner is designed to promote proper mixing of fuel and air for efficient combustion.

- Combustion systems may include burners, igniters, flame stabilizers, and other components that ensure reliable ignition and sustained combustion.

- Advanced combustion systems, such as low NOx burners, promote efficient fuel combustion while minimizing nitrogen oxide (NOx) emissions.

5. Combustion Control:

- Combustion control systems monitor and regulate various parameters, including fuel flow, air flow, and combustion chamber conditions.

- Control systems use sensors, feedback loops, and control algorithms to maintain optimal combustion conditions, maximizing efficiency and ensuring safe operation.

- Advanced control systems may include technologies like oxygen trim control, which continuously adjusts the air-to-fuel ratio based on real-time measurements.

6. Emissions Control:

- Emissions control technologies, such as flue gas treatment systems, help reduce air pollutants generated during combustion, including particulate matter, sulfur dioxide (SO_2), and nitrogen oxides (NOx).

- Flue gas treatment systems may include electrostatic precipitators, bag filters, selective catalytic reduction (SCR), or selective non-catalytic reduction (SNCR) systems, depending on the specific emissions requirements and regulations.

7. Efficiency Optimization:

- Efficient combustion systems and control strategies improve boiler efficiency by maximizing heat transfer, minimizing excess air, and reducing energy losses.

- Regular maintenance, combustion tuning, and performance monitoring are essential for sustaining high boiler efficiency and minimizing fuel consumption.

C. Water Treatment And Boiler Feedwater Systems

Water treatment and boiler feedwater systems play a vital role in maintaining efficient and reliable boiler operation. These systems are responsible for treating raw water to remove impurities and provide high-quality feedwater to the boiler. Here are the key aspects related to water treatment and boiler feedwater systems:

1. Water Source and Pretreatment:

- The water used for boiler feedwater can come from various sources such as surface water, groundwater, or municipal water supply.

- Pretreatment processes are employed to remove impurities, suspended solids, dissolved minerals, and other contaminants from the raw water.

- Common pretreatment methods include filtration, sedimentation, coagulation, and chemical dosing to ensure the water quality meets boiler requirements.

2. Deaeration:

- Deaeration is the process of removing dissolved oxygen from the feedwater to prevent corrosion and oxygen pitting in the boiler system.

- Deaerators use steam to heat and strip oxygen from the water, along with mechanical and chemical methods to enhance the oxygen removal efficiency.

3. Water Softening and Demineralization:

- Water softening is employed to reduce the hardness of water by removing calcium and magnesium ions, which can lead to scale formation.

- Demineralization processes, such as ion exchange or reverse osmosis, remove dissolved minerals, salts, and impurities to produce high-purity water for boiler feed.

4. Chemical Treatment:

- Chemical treatment involves the addition of chemicals to the boiler feedwater to control pH levels, prevent scale formation, and inhibit corrosion.

- Common chemicals used in boiler water treatment include oxygen scavengers, alkalinity builders, pH adjusters, scale inhibitors, and corrosion inhibitors.

- Proper dosage and monitoring of chemicals are crucial to maintain water quality and protect the boiler system from damage.

5. Feedwater Storage and Feed Pumps:

- Feedwater storage tanks provide a reservoir of treated water to ensure a continuous supply of feedwater to the boiler, especially during peak demands.

- Feed pumps are used to transfer the treated water from the storage tank to the boiler at the required pressure.

- The pumps should be properly sized and maintained to ensure reliable and efficient operation.

6. Feedwater Control and Monitoring:

- Feedwater control systems regulate the flow rate, pressure, and temperature of the feedwater to match the boiler's requirements.

- Control valves, flow meters, pressure sensors, and temperature sensors are utilized to monitor and adjust the feedwater conditions.

- Proper control and monitoring ensure the optimal supply of feedwater to the boiler, maintaining desired operating parameters.

7. Condensate Recovery:

- Condensate recovery systems recover and reuse the condensate produced from the steam generated by the boiler.

- Condensate is treated and returned to the boiler, reducing the need for fresh feedwater and saving energy.

- Condensate recovery systems typically include condensate pumps, heat exchangers, and treatment equipment.

D. Boiler Control Systems And Instrumentation

Boiler control systems and instrumentation are essential components for monitoring and regulating various parameters to ensure safe, efficient, and reliable boiler operation. These systems provide real-time data, control mechanisms, and safety interlocks to optimize combustion, maintain desired operating conditions, and protect the boiler and personnel. Here are key aspects related to boiler control systems and instrumentation:

1. Control Loops:
- Boiler control systems consist of various control loops that monitor and regulate different parameters. Common control loops include fuel flow, air flow, water level,

pressure, temperature, and combustion control.

- Each control loop consists of sensors to measure the parameter, a controller to process the data, and actuators to adjust the control elements (e.g., fuel valves, air dampers, feedwater pumps) based on the controller's output.

2. Sensors and Transmitters:

- Sensors and transmitters are used to measure and monitor critical parameters. Common sensors in boiler instrumentation include pressure sensors, temperature sensors, flow sensors, level sensors, oxygen analyzers, and combustion analyzers.

- The sensors provide real-time data to the control system, allowing for accurate monitoring and control of the boiler's operating conditions.

3. Programmable Logic Controllers (PLCs):

   - PLCs are widely used in boiler control systems for their ability to process data, execute control algorithms, and coordinate multiple control loops.

   - PLCs provide a centralized control platform that integrates various input/output (I/O) modules, allowing for efficient communication with sensors, actuators, and other control devices.

   - PLC-based control systems offer flexibility, scalability, and advanced functionality for boiler control and automation.

4. Human-Machine Interface (HMI):

   - HMIs provide an interface for operators to interact with the boiler control system. They display real-time data, control parameters, and alarms, allowing operators to monitor and adjust the boiler operation.

- HMIs typically include graphical representations of the boiler, trend charts, alarm notifications, and control buttons for manual operation.

5. Safety Interlocks and Alarms:

- Boiler control systems incorporate safety interlocks and alarms to protect against hazardous conditions and equipment malfunctions.

- Safety interlocks ensure that certain conditions are met before specific operations can be performed. For example, a low water level interlock may prevent the burner from operating if the water level falls below a critical point.

- Alarms provide visual or audible notifications to alert operators of abnormal conditions, equipment failures, or deviations from desired parameters.

6. Control Strategies:

  - Boiler control strategies define the algorithms and logic used to regulate the boiler's operation. They may involve proportional-integral-derivative (PID) control, feedforward control, cascade control, or advanced control techniques.

  - Control strategies aim to maintain stable and efficient boiler operation by continuously adjusting control elements based on sensor feedback and predefined setpoints.

7. Data Logging and Monitoring:

  - Boiler control systems often include data logging and monitoring capabilities. Historical data can be logged and analyzed to evaluate boiler performance, diagnose issues, and optimize operation.

- Remote monitoring systems enable operators to access real-time data and control the boiler operation from a central control room or through web-based interfaces.

E. Boiler Safety Devices And Regulations

Boiler safety devices and regulations are designed to ensure the safe operation of boilers, protect personnel, and prevent accidents or hazardous conditions. These devices and regulations aim to maintain proper pressure, temperature, water level, and fuel supply within safe limits. Here are some key safety devices and regulations associated with boilers:

1. Pressure Relief Valve (PRV) or Safety Valve:

- The PRV or safety valve is a crucial safety device that automatically releases excess pressure from the boiler to prevent over-pressurization and potential explosions.

- It is set to open at a specific pressure limit, known as the set pressure, and discharge steam or hot water to a safe location when the boiler pressure exceeds this limit.

- PRVs should be properly sized, installed, and tested periodically to ensure their reliability.

2. Low Water Cutoff (LWCO) and Water Level Controls:

- LWCO devices monitor the water level inside the boiler and shut off the fuel supply or activate an alarm if the water level falls below a safe operating level.

- Water level controls help maintain the desired water level within the boiler, preventing dry firing and damage to the boiler and its components.

- Proper calibration and regular maintenance of LWCO devices and water level controls are essential for reliable operation.

3. Flame Safeguard Systems:

- Flame safeguard systems monitor and control the ignition and flame stability within the boiler combustion chamber.

- These systems typically include flame detectors that ensure the presence of a flame during burner operation and initiate safety measures if the flame is lost.

- Flame safeguard systems provide a critical safety function by preventing uncontrolled fuel flow in the absence of a flame.

4. Fuel Train Safety Devices:

- Fuel trains include various safety devices to regulate and control the flow of fuel to the burner.

- Common safety devices in fuel trains include fuel pressure switches, fuel shut-off valves, and flame scanners that monitor the combustion process and ensure safe fuel supply.

- Regular inspection, testing, and maintenance of fuel train safety devices are necessary to ensure their proper functioning.

5. Boiler and Pressure Vessel Codes and Regulations:

- Boiler safety is governed by codes and regulations established by national and local authorities, such as the American Society of Mechanical Engineers (ASME) Boiler and Pressure Vessel Code.

- These codes provide guidelines for the design, construction, installation, and operation of boilers to ensure their safety and compliance with industry standards.

- Compliance with applicable codes and regulations is necessary for obtaining necessary permits, certifications, and ensuring safe boiler operation.

6. Regular Inspections and Maintenance:

- Regular inspections and maintenance of boilers are crucial for identifying potential safety hazards, verifying the integrity of safety devices, and addressing any issues promptly.

- Qualified personnel, such as boiler inspectors or certified technicians, should conduct routine inspections and maintenance in accordance with regulatory requirements and manufacturer guidelines.

7. Training and Operator Competence:

- Proper training and competence of boiler operators are essential for safe operation and adherence to safety procedures.

- Operators should receive training on boiler operations, safety protocols, emergency procedures, and the proper use of safety devices.

- Ongoing training and refresher courses ensure that operators stay updated with the latest safety practices and regulatory requirements.

F. Emergency Procedures And Troubleshooting

Emergency procedures and troubleshooting are crucial aspects of boiler operation to ensure the safety of personnel and mitigate potential risks. Being prepared for emergencies and having a systematic approach to troubleshooting can help address issues promptly and minimize downtime. Here are key considerations for emergency procedures and troubleshooting in boiler operation:

Emergency Procedures:

1. Emergency Shutdown:
 - In the event of a critical situation, such as excessive pressure, water level abnormalities, or fuel supply issues, the boiler should be shut down immediately.

- Emergency shutdown procedures should be clearly defined, communicated, and understood by all operators. These procedures typically involve shutting off fuel supply, isolating electrical power, and activating safety devices.

2. Evacuation and Personnel Safety:

- If the emergency poses a risk to personnel or the surrounding environment, appropriate evacuation procedures should be followed.

- Personnel should be trained on evacuation routes, assembly points, and emergency communication protocols to ensure their safety during emergencies.

3. Communication and Notification:

- Establish a clear communication plan to promptly inform relevant personnel, supervisors, and emergency response teams about the emergency situation.

- Provide clear instructions on who to notify, what information to convey, and how to coordinate the response efforts.

4. Emergency Services:

- In severe emergencies, it may be necessary to contact emergency services, such as the fire department or emergency medical services, for assistance and support.

Troubleshooting:

1. Identify and Assess the Problem:

- When troubleshooting boiler issues, start by identifying the specific problem or symptom. This could include abnormal pressure, temperature fluctuations, water leaks, unusual noises, or burner malfunctions.

- Gather information about the operating conditions, recent changes, and any alarms or warning indicators that may provide clues about the issue.

2. Safety Precautions:

- Before starting any troubleshooting activities, ensure that appropriate safety precautions are taken, such as wearing personal protective equipment (PPE) and isolating power sources.

3. Refer to Operating Manuals and Documentation:

- Review the boiler's operating manuals, technical documentation, and manufacturer guidelines to understand the system's design, components, and recommended troubleshooting procedures.
- These resources often include troubleshooting flowcharts, diagnostic

procedures, and common problem-solution scenarios.

4. Systematic Approach:

- Follow a systematic approach to troubleshooting by checking the various subsystems, components, and control settings of the boiler.
- Start with the simplest and most common issues, such as checking fuel supply, water level, and power supply, before moving on to more complex troubleshooting steps.

5. Data Analysis and Monitoring:

- Utilize available data, such as pressure, temperature, and flow measurements, to identify trends, patterns, or deviations from normal operation.
- Monitor and analyze real-time data from control systems or instrumentation to pinpoint potential causes of the problem.

Chapter 4: Boiler Maintenance And Inspection

A. Routine Boiler Maintenance Tasks

Routine boiler maintenance is essential for ensuring the efficient and reliable operation of the boiler, extending its lifespan, and minimizing the risk of breakdowns. Here are some key routine maintenance tasks for boilers:

1. Regular Inspections:
   - Conduct visual inspections of the boiler and its components to identify any signs of leaks, corrosion, or physical damage.
   - Inspect the burner, fuel and air supply systems, combustion chamber, heat exchanger, flue gas path, and controls for any abnormalities or wear.

- Inspect safety devices, such as pressure relief valves, low water cutoffs, and flame safeguard systems, to ensure they are in proper working condition.

2. Combustion System Maintenance:

- Clean and inspect the burner, nozzles, and flame sensors to ensure proper fuel combustion and optimize burner efficiency.
- Check and adjust the fuel-to-air ratio to achieve efficient combustion and minimize emissions.
- Clean or replace fuel filters, strainers, and air filters regularly to maintain clean and unobstructed fuel and air supply.

3. Water Treatment and Boiler Feed System Maintenance:

- Monitor and maintain proper chemical levels and dosing for water treatment to prevent scale, corrosion, and deposits.

- Inspect and clean the water level controls, feedwater pumps, and feedwater piping to ensure smooth operation and adequate water supply.

- Regularly test and treat boiler water to maintain optimal chemical balance and prevent scale formation.

4. Cleaning and Maintenance of Heat Exchanger Surfaces:

- Regularly clean heat exchanger surfaces, including tubes, flue gas passages, and soot blowers, to remove accumulated deposits and improve heat transfer efficiency.

- Use appropriate cleaning methods, such as brushing, scraping, or chemical cleaning, based on the type of deposits and the recommendations provided by the boiler manufacturer.

5. Safety Device Testing and Calibration:

- Test and verify the operation of safety devices, including pressure relief valves, low water cutoffs, flame detectors, and temperature/pressure controls.

- Follow manufacturer guidelines and regulatory requirements for testing frequencies and procedures.

- Ensure that safety devices are calibrated correctly and adjusted as necessary to maintain their accuracy and reliability.

6. Control System Calibration and Testing:

- Periodically calibrate and test the control system components, such as sensors, transmitters, control valves, and actuators.

- Verify that control setpoints, interlocks, and safety shutdowns are functioning properly.

- Conduct functional tests of the control system to ensure proper sequencing and operation of various boiler functions.

7. Inspection and Cleaning of Exhaust Systems:

- Inspect and clean the exhaust system, including flue gas ducts, stacks, and draft controls, to prevent obstructions and maintain proper draft conditions.
- Remove soot, ash, and other debris that may accumulate in the exhaust system.

8. Documentation and Maintenance Logs:

- Maintain detailed maintenance records, including inspection reports, maintenance logs, and equipment history.
- Document any repairs, replacements, or significant findings during routine maintenance activities.

B. Boiler Cleaning And Inspections

Boiler cleaning and inspections are essential maintenance tasks that help ensure the efficient and safe operation of boilers. Regular cleaning and inspections help prevent the accumulation of deposits, corrosion, and other issues that can adversely affect boiler performance. Here are the key aspects of boiler cleaning and inspections:

1. Internal Boiler Cleaning:

   - Boiler cleaning involves removing deposits, such as scale, soot, and ash, from the internal surfaces of the boiler, including the heat exchanger, tubes, and combustion chamber.

   - The cleaning method depends on the type and severity of the deposits. Common methods include brushing, water washing, chemical cleaning, or a combination of these techniques.

- Follow manufacturer recommendations and guidelines for the specific cleaning procedures and chemicals to use.

- Pay special attention to areas prone to high deposit accumulation, such as water-side surfaces, flue gas passages, and fire-side surfaces.

2. Sootblowing:

- Sootblowing is a cleaning method specifically used for removing soot and ash deposits from boiler surfaces, particularly in the heat exchanger and flue gas passages.

- Sootblowers are devices that emit steam or compressed air to dislodge and remove soot and ash.

- Regular sootblowing helps maintain heat transfer efficiency, prevent blockages, and minimize the risk of fires caused by excessive soot accumulation.

3. External Boiler Cleaning:

- External cleaning involves cleaning the exterior surfaces of the boiler, such as the casing, panels, and insulation.

- Remove dirt, dust, and debris that may accumulate on the external surfaces to maintain a clean and safe working environment.

- Ensure that cleaning methods and materials used do not damage or degrade the boiler's external components.

4. Boiler Inspections:

- Regular inspections of the boiler are essential for identifying potential issues, assessing the condition of components, and ensuring compliance with safety standards.

- Inspections may include visual inspections, non-destructive testing (NDT) techniques, and equipment-specific inspections.

- Inspect the boiler's internal and external surfaces, including the combustion chamber, heat exchanger, tubes, refractory, insulation, safety devices, and piping.

- Check for signs of corrosion, leaks, cracks, erosion, abnormal wear, or any other damage that may affect the boiler's integrity and performance.

- Conduct thickness measurements on critical components, such as tubes and pressure vessel walls, to monitor wear and determine if any sections require replacement.

5. Boiler Certification and Regulatory Inspections:

- Boilers are subject to periodic inspections by regulatory authorities to ensure compliance with safety standards and applicable codes.

- Consult local regulations and comply with inspection intervals and requirements set forth by the relevant authorities.

- Maintain proper documentation and certification records to demonstrate compliance with regulatory inspections.

6. Safety Considerations:

- Prioritize safety during boiler cleaning and inspections. Follow lockout/tagout procedures and ensure that the boiler is isolated from energy sources before starting any cleaning or inspection activities.

- Use appropriate personal protective equipment (PPE) as recommended by safety guidelines.

- Adhere to safety protocols when working at heights or in confined spaces, if applicable.

C. Boiler Tube Failure Mechanisms And Prevention

Boiler tube failures can significantly impact the operation and efficiency of a boiler system. Understanding the various failure mechanisms and implementing preventive measures can help mitigate the risk of tube failures. Here are common boiler tube failure mechanisms and prevention strategies:

1. Corrosion:

- Corrosion is a major cause of boiler tube failures. It can occur due to various factors, including water chemistry, dissolved gases, and temperature differentials.

- Preventive measures include proper water treatment to control pH levels, dissolved oxygen, and chemical concentrations. Regular monitoring and adjustment of water chemistry parameters are crucial.

- Utilize corrosion inhibitors, oxygen scavengers, and proper water treatment chemicals to minimize corrosion potential.

- Implement protective coatings or linings on vulnerable surfaces to provide an additional barrier against corrosion.

2. Erosion and Erosion-Corrosion:

- Erosion and erosion-corrosion result from the high-velocity flow of combustion gases or water droplets against boiler tubes.

- Employ measures such as installing erosion-resistant materials, baffles, or flow modifiers to redirect or reduce the impact of high-velocity flows.

- Optimize burner and fuel nozzle designs to minimize excessive gas velocities and reduce erosion potential.

- Regularly inspect and maintain boiler internals, such as baffles and tube supports, to prevent localized erosion.

3. Thermal Fatigue:

- Thermal fatigue occurs due to repeated heating and cooling cycles, causing stress and fatigue in the boiler tubes.

- Avoid rapid temperature changes and thermal shock by implementing proper warm-up and cool-down procedures during boiler start-up and shutdown.

- Maintain a stable and uniform temperature profile across the boiler to minimize temperature differentials and reduce thermal stress on tubes.

- Install insulation or refractory materials to provide thermal protection and reduce temperature gradients.

4. Overheating:

- Overheating can lead to tube failures due to excessive temperature and stress on the tubes.

- Ensure proper water circulation and flow rates to prevent localized hot spots and overheating.

- Implement adequate instrumentation and controls to monitor and maintain optimal operating temperatures.

- Regularly inspect and clean heat transfer surfaces to prevent the accumulation of deposits that can lead to overheating.

5. Mechanical Stress and Vibration:

- Mechanical stress and vibration can cause fatigue and failure of boiler tubes.

- Ensure proper alignment and support of tubes to minimize stress and vibration.

- Consider the use of expansion joints or flexible connections to accommodate thermal expansion and contraction.

- Perform regular inspections of tube supports, hangers, and attachments to identify and rectify any issues.

6. Water Treatment and Feedwater Quality:

- Poor water treatment and inadequate feedwater quality can contribute to tube failures.

- Maintain proper water chemistry through effective water treatment processes, including chemical dosing, filtration, and softening/demineralization.

- Monitor and control impurities, such as dissolved solids, hardness, and alkalinity, to prevent scale formation and corrosion.

- Ensure proper feedwater quality by filtering and removing suspended solids to prevent fouling and deposits on tube surfaces.

7. Regular Inspection and Maintenance:

- Conduct regular visual inspections, non-destructive testing (NDT), and thickness measurements to identify early signs of tube degradation.

- Address any identified issues promptly, such as replacing or repairing damaged tubes, repairing insulation, or modifying operational parameters as necessary.

- Implement a comprehensive preventive maintenance program, including cleaning, inspections, and testing, to ensure the overall health of the boiler system.

D. Boiler Efficiency Optimization Techniques

Optimizing boiler efficiency is crucial for reducing fuel consumption, minimizing operating costs, and maximizing the overall performance of a boiler system. Outlined below are several methods to enhance the efficiency of boilers:

1. Combustion Optimization:

- Ensure proper fuel combustion by maintaining the correct fuel-to-air ratio. Adjusting the combustion air flow rate and optimizing the burner settings can help achieve efficient combustion.
- Regularly inspect and clean burners, nozzles, and flame sensors to ensure optimal fuel atomization and flame characteristics.
- Utilize advanced combustion controls and technologies, such as oxygen trim

systems and flue gas analyzers, to continuously monitor and optimize combustion efficiency.

2. Heat Recovery Systems:

- Implement heat recovery systems to capture and utilize waste heat from flue gases, steam, or blowdown water.

- Install economizers to preheat the feedwater using waste heat from the flue gases before it enters the boiler.

- Utilize condensing economizers to recover heat from the flue gases by condensing the water vapor, particularly in high-efficiency boilers.

- Consider using heat exchangers to recover heat from blowdown water or process steam for preheating purposes.

3. Boiler and System Insulation:

- Properly insulate the boiler and distribution system components, such as pipes, valves, and fittings, to minimize heat losses.

- Insulate steam lines and condensate return lines to maintain the temperature and reduce energy losses during distribution.

- Insulate the boiler shell, doors, and access panels to minimize heat radiation losses.

4. Optimal Boiler Operation and Control:

- Operate the boiler at its design capacity and within the recommended load range to maximize efficiency.

- Implement advanced boiler control systems that optimize the combustion process, monitor operating parameters, and adjust boiler settings based on load demands.

- Regularly calibrate and maintain instrumentation and control devices to ensure accurate measurements and control.

- Optimize boiler start-up and shutdown procedures to minimize fuel consumption and thermal stress.

5. Water Treatment and Blowdown Optimization:

- Maintain proper water chemistry through effective water treatment to prevent scale formation and corrosion, which can reduce heat transfer efficiency.

- Optimize blowdown rates and schedules to control the concentration of dissolved solids in the boiler water.

- Utilize blowdown heat recovery systems to recover and utilize waste heat from the blowdown water.

6. Regular Maintenance and Cleaning:

- Implement a comprehensive maintenance program that includes regular cleaning of heat transfer surfaces, such as tubes and heat exchangers, to remove scale, soot, and other deposits.

- Clean or replace air and fuel filters regularly to ensure proper combustion and airflow.

- Inspect and maintain boiler components, such as burner nozzles, insulation, refractory, and seals, to ensure optimal performance.

7. Continuous Monitoring and Optimization:

- Utilize energy management systems and real-time monitoring tools to track boiler performance, energy consumption, and efficiency metrics.

- Regularly analyze and evaluate boiler performance data to identify areas for improvement and implement corrective measures.

8. Staff Training and Awareness:

- Provide comprehensive training to boiler operators and maintenance personnel on best practices for efficient boiler operation, maintenance, and troubleshooting.

- Foster a culture of energy efficiency and awareness among staff members to encourage proactive energy-saving behaviors.

E. Boiler Repair And Replacement Considerations

Boilers, like any mechanical equipment, may require repairs or eventual replacement due to various factors such as age, wear and tear, and efficiency concerns. Here are some considerations when deciding between boiler repair and replacement:

1. Boiler Age and Condition:

 - Consider the boiler's age. Older boilers may be less efficient and more prone to breakdowns.

 - Assess the overall condition of the boiler, including the heat exchanger, tubes, controls, and safety devices. If the boiler has significant corrosion, leaks, or structural issues, replacement may be more appropriate.

- Evaluate the boiler's maintenance history. If it has a consistent record of breakdowns and repairs, it may be more cost-effective to replace it with a newer, more reliable unit.

2. Efficiency and Energy Savings:

- Compare the energy efficiency ratings of the existing boiler with newer models. Advances in technology have resulted in more efficient boilers, which can lead to energy savings and lower operating costs.

- Calculate the potential energy savings over the lifespan of a new boiler to determine the cost-effectiveness of replacement.

3. Repair Costs vs. Replacement Costs:

- Assess the cost of repairs versus the cost of a new boiler. Consider the frequency and extent of repairs needed.

- Consult with qualified technicians or boiler service providers to obtain accurate estimates for repairs and replacement. Take into account not only the cost of the boiler unit itself but also installation and any necessary modifications to the system.

4. Availability of Parts and Technical Support:

- Consider the availability of replacement parts for the existing boiler. If the boiler is outdated or obsolete, finding suitable parts may be difficult or expensive.

- Evaluate the availability of technical support and expertise for the existing boiler. If it is a less common or specialized model, it may be challenging to find qualified technicians for repairs and maintenance.

5. Safety and Compliance:

- Ensure that the existing boiler meets current safety standards and regulations. If the boiler is outdated and non-compliant, replacement may be necessary to ensure the safety of the facility and occupants.

- Consider any changes in regulations or emission standards that may affect the operation of the boiler. Compliance with environmental regulations may warrant replacement with a more environmentally friendly unit.

6. Future Needs and Capacity:

- Evaluate the current and future heating requirements of the facility. If there are plans for expansion or changes in heating demands, consider whether the existing boiler can adequately meet those needs.

- Assess the capacity of the existing boiler. If it is undersized or oversized for the current load, replacement with a properly sized unit can optimize efficiency and performance.

7. Financial Considerations:

- Take into account available budget and financing options for repairs or replacement. Consider the potential return on investment (ROI) and payback period for a new, more efficient boiler.

- Explore any available incentives or rebates for energy-efficient boilers, which may help offset the upfront costs of replacement.

Chapter 5: Boiler Efficiency And Energy Conservation

A. Understanding Boiler Efficiency And Performance

Understanding boiler efficiency and performance is essential for evaluating the effectiveness and energy consumption of a boiler system. Below are the vital as well as key aspects to put into consideration:

1. Boiler Efficiency:

- Boiler efficiency refers to the ratio of useful energy output (e.g., heat transferred to the system) to the energy input (e.g., fuel combustion). It indicates how effectively the boiler converts fuel energy into useful heat.

- Boiler efficiency is commonly quantified as a percentage, with higher percentages indicating greater efficiency of the boiler.

- Various forms of boiler efficiency exist:

- Combustion Efficiency: It measures how effectively the fuel is burned and converted into heat. Factors affecting combustion efficiency include fuel quality, burner design, and proper air-to-fuel ratio.

- Thermal Efficiency: It measures how effectively the boiler transfers heat from the combustion process to the water or steam. Thermal efficiency is influenced by heat exchanger design, insulation, and flue gas temperature.

- Overall Efficiency: It represents the combined efficiency of both combustion and thermal efficiency. It considers all energy losses, including radiation losses and losses in heat transfer surfaces.

2. Factors Affecting Boiler Efficiency:

- Combustion Air: Proper air-to-fuel ratio is crucial for efficient combustion.

Insufficient or excessive combustion air can affect boiler efficiency and increase fuel consumption.

- Fuel Quality: The quality and characteristics of the fuel, such as moisture content, ash content, and heating value, can impact boiler performance.

- Heat Exchanger Design: The design and condition of heat exchangers, including tubes, baffles, and fins, influence heat transfer efficiency.

- Scale and Deposits: Accumulation of scale, soot, or other deposits on heat transfer surfaces reduces heat transfer efficiency and increases fuel consumption.

- Flue Gas Temperature: Lowering the flue gas temperature by implementing heat recovery systems, such as economizers or condensing systems, can improve boiler efficiency.

- Boiler Load: Efficiency may vary at different load levels. Operating the boiler closer to its rated capacity can optimize efficiency.

- Boiler Control and Optimization: Advanced control systems and optimization techniques can help maintain efficient boiler operation by adjusting firing rates, optimizing air-to-fuel ratios, and managing load fluctuations.

3. Performance Monitoring and Measurement:

- Regular monitoring and measurement of key performance indicators (KPIs) are essential for evaluating boiler efficiency and identifying potential areas for improvement.

- Common KPIs include fuel consumption, steam or hot water production, flue gas temperature, stack emissions, and operating hours.

- Utilize instrumentation and monitoring devices, such as flow meters, temperature sensors, and combustion analyzers, to gather accurate data on boiler performance.

- Establish a routine maintenance and inspection schedule to ensure the accuracy and reliability of measurement devices.

4. Benchmarking and Comparison:

- Benchmarking boiler efficiency against industry standards and similar boiler systems can provide insights into performance gaps and potential improvements.

- Utilize efficiency standards, such as ASME PTC 4.1 or ISO 9001, as benchmarks for evaluating and comparing boiler performance.

5. Efficiency Improvement Strategies:

- Implement energy-saving measures, such as insulation, heat recovery systems, and combustion optimization, to improve boiler efficiency.

- Regularly clean and maintain heat transfer surfaces to prevent the buildup of scale and deposits.

- Optimize boiler operation and load management based on the facility's heating demands.

- Invest in modern, high-efficiency boilers that meet energy efficiency standards and offer advanced control features.

B. Combustion Optimization And Air-To-Fuel Ratio Control

Combustion optimization and proper control of the air-to-fuel ratio are crucial for achieving efficient and clean combustion in a boiler system. Here's an overview of combustion optimization techniques and air-to-fuel ratio control:

1. Combustion Optimization:

 - Combustion optimization aims to achieve complete combustion of fuel while minimizing energy losses and emissions.

 - Proper combustion improves boiler efficiency, reduces fuel consumption, and minimizes the formation of harmful pollutants, such as carbon monoxide (CO) and nitrogen oxides (NOx).

 - Techniques for combustion optimization include:

- Burner Tuning: Adjusting burner settings, such as air registers, fuel pressure, and atomization, to optimize the fuel-air mixture and flame characteristics.

- Flame Monitoring: Using flame sensors or scanners to monitor the flame quality and stability, ensuring efficient combustion.

- Combustion Air Preheating: Preheating combustion air using waste heat or dedicated heat exchangers improves combustion efficiency.

- Flue Gas Recirculation (FGR): Introducing a portion of the flue gas back into the combustion air helps reduce flame temperature and lower NOx emissions.

- Advanced Burner Technologies: Utilizing advanced burner designs, such as low-NOx burners or staged combustion systems, to achieve more efficient and cleaner combustion.

2. Air-to-Fuel Ratio Control:

- The air-to-fuel ratio represents the ratio of combustion air (oxygen) to fuel for proper combustion.

- Maintaining the correct air-to-fuel ratio is essential for efficient combustion. Too much air (lean condition) or too little air (rich condition) can lead to energy losses, increased emissions, and reduced boiler performance.

- Control methods for air-to-fuel ratio include:

  - Oxygen Trim Systems: Utilizing oxygen sensors in the flue gas to measure the oxygen content and adjusting the combustion air supply accordingly. This ensures precise control of the air-to-fuel ratio and optimal combustion efficiency.

  - Flue Gas Analysis: Regularly analyzing the flue gas composition using gas analyzers to determine the combustion efficiency and adjust the air-to-fuel ratio as needed.

- Stoichiometric Control: Calculating the theoretical air requirement based on the fuel's chemical composition and adjusting the combustion air supply accordingly. This approach ensures the ideal air-to-fuel ratio for complete combustion.

- Burner Controls: Implementing advanced burner controls that monitor and adjust the air and fuel supply rates to maintain the desired air-to-fuel ratio.

- Variable Frequency Drives (VFDs): Using VFDs to control the airflow rate and match it to the boiler's load requirements, ensuring the optimal air-to-fuel ratio at different operating conditions.

C. Waste Heat Recovery And Heat Exchangers

Waste heat recovery (WHR) and heat exchangers play a crucial role in improving energy efficiency and reducing energy waste in various industrial processes. Here's an overview of waste heat recovery and the use of heat exchangers:

1. Waste Heat Recovery (WHR):

- Waste heat is generated as a byproduct of various industrial processes, such as combustion, exhaust gases, and hot process streams.

- Waste heat recovery involves capturing and utilizing this waste heat to generate additional useful energy or provide heat for other processes.

- WHR systems can significantly improve energy efficiency, reduce fuel consumption, and lower greenhouse gas emissions.

- Common sources of waste heat include flue gases from boilers, exhaust gases from engines and turbines, and hot process streams in manufacturing operations.

2. Heat Exchangers:

- Heat exchangers are devices designed to transfer heat from one fluid to another fluid without direct contact, allowing efficient heat exchange between the two streams.

- Heat exchangers play a critical role in waste heat recovery by facilitating the transfer of heat from the waste heat source to a fluid that can be utilized for another purpose.

- Different varieties of heat exchangers are available, including:

- Shell and Tube Heat Exchangers: These consist of a shell with multiple tubes inside. While one fluid circulates within the tubes, the other fluid traverses through the shell. The transfer of heat occurs from the hotter fluid to the colder fluid by means of the tube walls.

- Plate Heat Exchangers: These use a series of corrugated plates to create a large surface area for heat transfer between the two fluids.

- Finned Tube Heat Exchangers: These have extended surfaces (fins) on the tubes to enhance heat transfer efficiency.

- Air-to-Air Heat Exchangers: These transfer heat between two separate air streams, commonly used in ventilation systems or waste heat recovery from exhaust air.

3. Waste Heat Recovery Applications:

- Waste heat recovery can be applied in various industrial processes and systems, such as:

- Boiler Economizers: Heat exchangers installed in the flue gas path of boilers to preheat the boiler feedwater, reducing fuel consumption.

- Condensing Economizers: Heat exchangers that recover latent heat from the flue gases by condensing the water vapor, achieving higher overall efficiency.

- Heat Recovery Steam Generators (HRSG): Used in combined cycle power plants, HRSGs recover waste heat from gas turbines to produce steam for additional power generation.

- Organic Rankine Cycle (ORC) Systems: These utilize waste heat to generate electricity by using an organic working fluid with a lower boiling point than water.

- Heat Recovery from Process Streams: Heat exchangers can recover heat from hot process streams, such as exhaust gases, flue gases, or liquids, and transfer it to preheat incoming fluids or provide heat for other processes.

4. Benefits of Waste Heat Recovery and Heat Exchangers:

- Improved Energy Efficiency: Waste heat recovery reduces energy waste by utilizing otherwise wasted heat, resulting in reduced fuel consumption and operating costs.

- Environmental Impact: By recovering waste heat and improving energy efficiency, the use of heat exchangers contributes to the reduction of greenhouse gas emissions and environmental impact.

- Cost Savings: Waste heat recovery systems can lead to significant cost savings by reducing the need for additional fuel or energy sources.

- Process Optimization: Utilizing waste heat for other processes or preheating purposes can enhance process efficiency and overall plant performance.

D. Energy Conservation Techniques For Boilers

Energy conservation techniques for boilers are crucial for improving efficiency, reducing fuel consumption, and minimizing environmental impact. Here are some key techniques to consider:

1. Combustion Optimization:

- Proper combustion optimization, as discussed earlier, ensures efficient fuel utilization and reduces energy waste. It involves adjusting the air-to-fuel ratio, optimizing burner performance, and utilizing advanced combustion control systems.

2. Waste Heat Recovery:

- Waste heat recovery systems capture and utilize waste heat generated during the combustion process or from hot process streams.

- Install heat exchangers or economizers to recover heat from flue gases and preheat boiler feedwater or other process streams.

- Consider the use of technologies like heat recovery steam generators (HRSGs) or organic Rankine cycle (ORC) systems to generate additional power from waste heat.

3. Boiler Insulation:

- Proper insulation of boilers and associated equipment minimizes heat loss and improves energy efficiency.

- Insulate boiler surfaces, steam distribution pipes, valves, and fittings to reduce heat radiation and prevent energy losses.

- Insulating boiler rooms and steam traps can also help maintain optimal operating conditions.

4. Load Management:

- Optimize boiler operation based on actual heating demand to avoid unnecessary energy consumption.

- Use advanced control systems to adjust firing rates, fuel-to-air ratios, and other parameters based on load fluctuations.

- Implement strategies like load tracking, load sharing, or cascading to optimize the operation of multiple boilers in a system.

5. Water Treatment:

- Proper water treatment minimizes scale and deposits on heat transfer surfaces, improving heat transfer efficiency.

- Use water treatment methods such as softening, demineralization, or chemical treatment to maintain clean and efficient heat transfer surfaces.

6. Regular Maintenance and Cleaning:

- Implement a routine maintenance schedule to ensure optimal boiler performance and efficiency.

- Clean heat transfer surfaces, remove scale or deposits, and inspect combustion components regularly.

- Replace worn-out or inefficient parts to maintain peak performance.

7. Continuous Monitoring and Control:

- Install monitoring devices and control systems to track key performance indicators (KPIs) and optimize boiler operation.

- Monitor fuel consumption, flue gas temperature, stack emissions, and other relevant parameters to identify opportunities for improvement.

- Utilize advanced control technologies, such as variable frequency drives (VFDs) or advanced control algorithms, to optimize boiler performance and energy efficiency.

8. Staff Training and Awareness:

- Provide training and awareness programs to boiler operators and maintenance personnel on energy conservation practices and efficient boiler operation.

- Promote good operational practices, such as regular inspections, efficient startup and shutdown procedures, and awareness of energy-saving opportunities.

Chapter 6: Boiler Operations In Specialized Industries

A. Boilers In Power Generation Plants

Boilers play a crucial role in power generation plants by providing the necessary steam or hot water to drive turbines and generate electricity. Here is an overview of the role and types of boilers used in power generation plants:

1. Role of Boilers in Power Generation:

- Boilers are primary components in power plants that produce steam or hot water to drive steam turbines or generate heat for other power generation processes.

- The heat energy produced by boilers is typically derived from the combustion of fossil fuels, such as coal, natural gas, or oil. However, some power plants also use renewable energy sources, such as biomass or waste heat, to generate steam.

- Boilers are responsible for converting water into steam by transferring heat to the water through combustion or other heat sources. The high-pressure steam is then used to drive steam turbines, which generate electricity through mechanical rotation.

2. Types of Boilers in Power Generation Plants:

- Pulverized Coal (PC) Boilers: These boilers burn pulverized coal and are commonly used in coal-fired power plants. The coal is pulverized into a fine powder and blown into the combustion chamber, where it is ignited to generate high-temperature flue gases that transfer heat to the boiler water.

- Circulating Fluidized Bed (CFB) Boilers: CFB boilers are capable of burning a wide range of fuels, including coal, biomass, and waste materials. They use a fluidized bed of sand or limestone particles to suspend and combust the fuel, providing high combustion efficiency and low emissions.

- Natural Gas and Oil-Fired Boilers: These boilers burn natural gas or oil as the primary fuel source. They are commonly used in combined cycle power plants, where the waste heat from the gas turbine exhaust is recovered to generate additional steam or hot water.

- Biomass Boilers: Biomass boilers utilize organic materials, such as wood pellets, agricultural residues, or dedicated energy crops, as fuel sources. They are considered renewable energy options and are often used in dedicated biomass power plants.

- Waste Heat Recovery Boilers: These boilers recover waste heat from various sources, such as exhaust gases from gas turbines or process heat streams. The recovered heat is used to generate steam or hot water, improving overall energy efficiency in power plants.

- Nuclear Power Plant Boilers: Nuclear power plants use nuclear reactors to generate heat through nuclear fission. The heat is then transferred to the boiler water, producing steam that drives the turbines.

3. Boiler Components in Power Generation Plants:

- Furnace: The combustion chamber where fuel is burned to produce heat.

- Water Walls: Heat-absorbing tubes or panels that line the furnace walls, absorbing heat from the combustion process and generating steam.

- Superheater: A section of the boiler that further heats the steam to a higher temperature to enhance the efficiency of the power generation process.

- Economizer: A heat exchanger that preheats the boiler feedwater using waste heat from the flue gases, increasing overall efficiency.

- Air Preheater: Another heat exchanger that preheats the combustion air using waste heat from the flue gases, improving combustion efficiency.

B. Boilers In Chemical And Petrochemical Industries

Boilers play a vital role in the chemical and petrochemical industries, where they are used for a wide range of applications. Here is an overview of the use of boilers in these industries:

1. Steam Generation:

- One of the primary uses of boilers in chemical and petrochemical industries is steam generation. Steam is essential for various processes, such as heating, drying, distillation, and chemical reactions.

- Boilers produce high-pressure steam by combusting fuels or utilizing waste heat. This steam is used in processes like steam cracking, hydrogen production, solvent recovery, and steam stripping.

2. Heat Transfer:

- Boilers are employed to transfer heat to process fluids and provide thermal energy for heating or cooling operations.
- Heat transfer fluids, such as thermal oil or water/glycol mixtures, are heated by the boiler and circulated through heat exchangers to heat or cool other process streams.

3. Process Heating:

- Boilers are used to provide direct heat for various chemical reactions and process heating requirements.
- They supply hot water or steam for processes like polymerization, drying, evaporation, sterilization, and chemical synthesis.

4. Power Generation:

- Some chemical and petrochemical plants have co-generation facilities where they generate electricity in addition to producing steam.

- Boilers play a crucial role in these power generation systems by providing steam to drive turbines and generate electricity.

5. Waste Heat Recovery:

- Waste heat recovery boilers are used in chemical and petrochemical industries to recover and utilize waste heat from various sources.

- Waste heat from flue gases, process streams, or exhaust gases is captured and used to produce steam or hot water, reducing the overall energy consumption and improving efficiency.

6. Boiler Types and Configurations:

- Chemical and petrochemical industries use a variety of boiler types, including:

- Fire-Tube Boilers: These boilers have hot gases flowing through tubes, transferring heat to the surrounding water.

- Water-Tube Boilers: In these boilers, water flows through tubes while hot gases pass around them, promoting efficient heat transfer.

- Package Boilers: Compact, pre-assembled boilers that can be easily transported and installed at the plant site.

- Waste Heat Recovery Boilers: These boilers recover waste heat from various sources, such as flue gases or process streams, to generate steam or hot water.

- High-Pressure Boilers: Used for processes requiring high-pressure steam, such as steam cracking or power generation.

- Utility Boilers: Large boilers that provide steam and heat for various plant processes, often operating at high temperatures and pressures.

C. Boilers In Food Processing And Pharmaceutical Industries

Boilers play a vital role in the food processing and pharmaceutical industries, where strict hygiene, safety, and quality standards are crucial. Here is an overview of the use of boilers in these industries:

1. Steam Generation:
   - Boilers are widely used in food processing and pharmaceutical industries to generate steam for various applications.
   - Steam is used for cooking, sterilization, pasteurization, cleaning, and heating processes in food processing facilities.
   - In the pharmaceutical industry, steam is used for sterilizing equipment, cleaning and sanitizing, as well as in certain manufacturing processes.

2. Heat Transfer:

- Boilers are utilized for heat transfer in food processing and pharmaceutical industries.

- Heat transfer fluids, such as thermal oil or water/glycol mixtures, are heated by the boiler and circulated through heat exchangers for heating or cooling processes.

- Heat exchangers are used for processes like heating or cooling product streams, maintaining precise temperature control, or condensing vapors.

3. Sterilization and Sanitization:

- Boilers are crucial for sterilizing equipment, packaging materials, and containers in food processing and pharmaceutical industries.

- Steam from boilers is used in autoclaves or sterilization tunnels to kill microorganisms and ensure product safety and shelf life.

- Clean steam, produced from high-purity water, is utilized for direct contact with products, ensuring the highest level of sterilization and quality.

4. Hot Water Supply:

- Boilers provide hot water for various applications in food processing and pharmaceutical industries.
- Hot water is used for cleaning and sanitizing equipment, as well as for general facility heating and domestic use.

5. Process Heating and Cooking:

- Boilers are used for process heating and cooking in food processing facilities.
- They supply hot water or steam for processes like blanching, cooking, baking, frying, drying, and evaporation.

6. Boiling and Distillation:

- Boilers are employed in food processing and pharmaceutical industries for boiling and distillation processes.

- They generate steam to heat liquids for boiling, concentration, extraction, or separation of desired components.

7. Compliance with Industry Standards:

- Boilers used in food processing and pharmaceutical industries must comply with stringent hygiene, safety, and quality standards.

- Materials of construction and design should meet industry regulations to ensure compatibility with food and pharmaceutical products.

- Regular maintenance, inspection, and cleaning are necessary to prevent contamination and ensure safe and reliable boiler operation.

8. Boiler Types and Configurations:

- The choice of boiler types in food processing and pharmaceutical industries depends on specific requirements, such as steam pressure, capacity, and process needs.

- Common boiler types include fire-tube boilers, water-tube boilers, and electric boilers.

- Hygienic design considerations, such as smooth surfaces, clean-in-place (CIP) systems, and stainless-steel construction, are essential for food and pharmaceutical applications.

D. Boilers In Paper And Pulp Mills

Boilers are essential components in paper and pulp mills, providing the necessary steam and heat for various processes involved in the production of paper and pulp products. Here is an overview of the use of boilers in paper and pulp mills:

1. Steam Generation:
  - Boilers are primarily used in paper and pulp mills for steam generation. Steam is a crucial utility in these industries and is utilized for various purposes throughout the production process.
  - Steam is used for heating, drying, and evaporating water from the pulp during the papermaking process.
  - It is also used for powering steam turbines, which generate electricity for internal use or for sale to the grid.

2. Recovery Boilers:

- Paper and pulp mills often employ recovery boilers to recover and utilize chemicals and energy from the pulping process.

- Recovery boilers burn black liquor, a byproduct of the pulping process, to generate steam and recover valuable chemicals, such as lignin and inorganic compounds.

- The steam produced from recovery boilers is used for multiple purposes, including heating, drying, and power generation.

3. Biomass Boilers:

- Some paper and pulp mills use biomass boilers as a renewable and sustainable alternative to traditional fossil fuel boilers.

- Biomass boilers burn biomass materials, such as wood chips, bark, and sawdust, to produce steam for various processes.

- Biomass boilers contribute to reducing the environmental impact of the paper and pulp industry by utilizing biomass waste from the production process or dedicated energy crops.

4. Lime Kiln:

- Paper and pulp mills often have lime kilns as a part of their production process.

- Lime kilns are used to convert calcium carbonate (lime mud) into calcium oxide (lime) through a calcination process.

- Boilers provide heat for the lime kiln, supplying the necessary energy to drive the calcination reaction and produce lime for use in the pulp production process.

5. Black Liquor Evaporation:

- Black liquor, a byproduct of the pulp production process, is concentrated through evaporation to recover chemicals and produce concentrated black liquor.

- Boilers supply the necessary steam for black liquor evaporation, which helps in the recovery of chemicals and reduces waste.

6. Heat Transfer:

- Boilers are used for heat transfer in various parts of the paper and pulp production process.

- Heat transfer fluids, such as thermal oil or steam, are circulated through heat exchangers to transfer heat to different process streams, such as drying cylinders or paper machines.

7. Power Generation:

- Paper and pulp mills often have power generation facilities where boilers play a crucial role.

- The steam produced by boilers can be used to drive steam turbines, generating electricity for internal use within the mill or for sale to the grid.

8. Boiler Types and Configurations:

- The choice of boiler types in paper and pulp mills depends on factors such as steam pressure, capacity, fuel availability, and specific process requirements.

- Common boiler types include recovery boilers, biomass boilers, black liquor boilers, and auxiliary boilers for heat transfer and power generation.

E. Boilers In HVAC Systems

Boilers are an integral part of HVAC (Heating, Ventilation, and Air Conditioning) systems, providing efficient heating for residential, commercial, and industrial buildings. Here is an overview of the use of boilers in HVAC systems:

1. Space Heating:

- Boilers are primarily used for space heating in HVAC systems. They provide heat to raise the temperature of air or water used to warm the indoor environment.

- Boilers can heat water or produce steam, which is then circulated through radiators, baseboard heaters, or underfloor heating systems to provide warmth to different areas of a building.

2. Hot Water Supply:

- Boilers are also responsible for providing hot water for domestic use in HVAC systems.

- Hot water generated by boilers is stored in tanks or circulated through pipes to deliver hot water for bathing, washing, and other household or commercial needs.

3. Hydronic Heating Systems:

- Boilers are commonly used in hydronic heating systems, where hot water or steam is circulated through pipes to transfer heat to various areas within a building.

- Hydronic systems can be used for space heating, radiant floor heating, or as a heat source for air handling units.

4. Radiant Heating:

- Boilers play a crucial role in radiant heating systems, where heat is emitted from a heated surface, such as radiators or radiant panels.

- Water or steam from the boiler flows through these surfaces, releasing heat energy and warming the surrounding space.

5. Absorption Cooling:

- In some HVAC systems, boilers are used in conjunction with absorption chillers to provide both heating and cooling.

- Absorption chillers use heat energy from the boiler to drive the cooling process, providing chilled water for air conditioning purposes.

6. Boiler Types and Configurations:

- Boilers used in HVAC systems can be of various types, including:

- Fire-Tube Boilers: These boilers have hot gases flowing through tubes, transferring heat to water surrounding the tubes.

- Water-Tube Boilers: In these boilers, water flows through tubes while hot gases pass around them, promoting efficient heat transfer.

- Condensing Boilers: Condensing boilers extract additional heat from the flue gases, maximizing efficiency by condensing the water vapor in the exhaust.

- Combination Boilers: Combination boilers provide both space heating and hot water supply from a single unit, eliminating the need for separate systems.

- Electric Boilers: Electric boilers use electricity as the energy source to generate heat, suitable for applications where other fuel sources are not available.

Chapter 7: Environmental Considerations

A. Emissions Control and Environmental Regulations

Emissions control and adherence to environmental regulations are critical aspects of boiler operation to minimize the impact on air quality and protect the environment. Here are some key considerations regarding emissions control and environmental regulations for boilers:

1. Emissions Monitoring:

- Boilers are subject to emissions monitoring to measure and track pollutant emissions such as nitrogen oxides (NOx), sulfur oxides (SOx), carbon monoxide (CO), particulate matter (PM), and volatile organic compounds (VOCs).

- Emission monitoring systems, such as continuous emissions monitoring systems (CEMS), are used to measure and report emissions levels to ensure compliance with regulatory standards.

2. Combustion Control:

- Proper combustion control techniques and technologies are employed to optimize fuel combustion and minimize emissions.

- Combustion control systems monitor and adjust key parameters such as air-to-fuel ratio, burner performance, and combustion efficiency to achieve cleaner and more efficient combustion.

3. Flue Gas Treatment:

- Flue gas treatment systems, such as flue gas desulfurization (FGD) and selective catalytic reduction (SCR), are utilized to remove or reduce pollutants from flue gases before they are released into the atmosphere.

- FGD systems remove sulfur dioxide (SO_2) from flue gases, while SCR systems reduce nitrogen oxide (NOx) emissions by converting them into nitrogen and water vapor.

4. Particulate Matter Control:

- Particulate matter (PM) emissions from boilers can be controlled through the use of electrostatic precipitators (ESPs), bag filters, or other particulate control devices.

- These devices capture and remove fine particles from flue gases, preventing their release into the environment.

5. Ash Disposal:

- Proper disposal and management of boiler ash, including bottom ash and fly ash, are essential to prevent contamination of soil, water, and air.

- Ash disposal methods must comply with environmental regulations, and in some cases, ash is recycled or used in other applications such as construction materials.

6. Environmental Regulations:

- Boilers are subject to various environmental regulations that set emission limits, dictate operating standards, and require compliance with specific monitoring and reporting requirements.

- These regulations can vary by region and may include national, state/provincial, and local standards that boilers must adhere to.

7. Environmental Certifications:

- Boilers may undergo certification processes to demonstrate compliance with specific environmental standards and regulations.

- Examples include certifications such as the U.S. Environmental Protection Agency's ENERGY STAR certification for energy efficiency or certifications for compliance with specific emission limits, such as the European Union's Industrial Emissions Directive (IED).

8. Renewable Energy Integration:

- Integration of renewable energy sources, such as biomass or solar thermal systems, can reduce the environmental impact of boilers by utilizing sustainable and low-carbon fuel sources.

- Renewable energy integration can help meet emission reduction goals and contribute to a cleaner and more sustainable energy mix.

B. Boiler Efficiency And Carbon Footprint Reduction

Improving boiler efficiency and reducing the carbon footprint are key objectives in promoting sustainable and environmentally friendly boiler operation. Here are some strategies to enhance boiler efficiency and minimize carbon emissions:

1. Combustion Optimization:

- Optimize the combustion process by maintaining the proper air-to-fuel ratio. Excess air can waste energy and increase emissions, while insufficient air can lead to incomplete combustion and higher levels of pollutants.

- Use advanced burner technology and control systems to achieve more precise and efficient combustion.

2. Heat Recovery Systems:

- Implement heat recovery systems, such as economizers and condensing heat exchangers, to capture waste heat from flue gases and use it to preheat boiler feedwater or other process streams.

- Heat recovery systems significantly improve overall boiler efficiency by utilizing waste heat that would otherwise be lost.

3. Insulation and Boiler Tune-Ups:

- Ensure proper insulation of the boiler and associated piping to minimize heat loss.

- Regular boiler tune-ups, including combustion analysis, equipment cleaning, and maintenance, help optimize performance and prevent efficiency losses.

4. Energy Management Systems:

  - Employ advanced energy management systems to monitor and control boiler operation, enabling better control of boiler parameters and optimizing energy consumption.

  - Utilize automated controls, sensors, and feedback mechanisms to adjust boiler operation based on load demand and maximize efficiency.

5. Efficient Boiler Design:

  - Select boilers with high-efficiency ratings and advanced design features, such as enhanced heat transfer surfaces, improved insulation, and optimized combustion chambers.

  - Consider the use of condensing boilers that recover additional heat from flue gases by condensing the water vapor.

6. Fuel Switching and Alternative Fuels:

- Explore the use of cleaner and more sustainable fuels, such as natural gas or biofuels, as alternatives to fossil fuels.

- Fuel switching can significantly reduce carbon emissions and improve the environmental profile of boiler operation.

7. Regular Maintenance and Upgrades:

- Implement regular maintenance programs to ensure boilers operate at peak efficiency.

- Regularly inspect and clean boiler components, optimize burner performance, and replace worn-out parts.

- Consider upgrading older boilers with newer, more efficient models or retrofitting existing boilers with efficiency-enhancing technologies.

8. Renewable Energy Integration:

- Integrate renewable energy sources, such as solar thermal or biomass, into boiler systems to reduce reliance on fossil fuels and lower carbon emissions.

- Hybrid systems that combine boilers with renewable energy sources can optimize energy use and minimize environmental impact.

9. Carbon Offsetting and Emissions Reduction Programs:

- Consider participating in carbon offset programs or implementing emissions reduction initiatives to compensate for unavoidable emissions.

- Invest in projects that offset carbon emissions by promoting renewable energy, reforestation, or energy efficiency.

C. Alternative and Renewable Energy Sources for Boilers

Boilers traditionally rely on fossil fuels such as coal, oil, and natural gas as their primary energy sources. However, with the increasing emphasis on sustainability and the transition to cleaner energy sources, alternative and renewable energy options for boilers have gained prominence. Here are some alternative and renewable energy sources that can be used in boilers:

1. Biomass:

- Biomass refers to organic materials derived from plants, such as wood pellets, agricultural residues, or dedicated energy crops.
- Biomass boilers use these biomass fuels to generate heat and steam, offering a

renewable and carbon-neutral alternative to fossil fuels.

- Biomass combustion releases CO_2, but it is considered carbon-neutral since the carbon emitted is offset by the carbon absorbed during the growth of the biomass feedstock.

2. Biogas:

- Biogas is produced through the anaerobic digestion of organic waste materials, such as agricultural waste, food waste, or sewage.

- Biogas can be used directly as a fuel in boilers, or it can be upgraded to biomethane, which has similar properties to natural gas.

- Biogas is a renewable energy source that reduces greenhouse gas emissions and helps in waste management.

3. Waste-to-Energy (WtE):

- Waste-to-Energy systems convert municipal solid waste (MSW) or industrial waste into heat and electricity.

- In WtE facilities, waste is incinerated, and the heat generated is used to produce steam for boiler applications.

- WtE offers a sustainable solution for waste management while simultaneously generating energy.

4. Solar Thermal Energy:

- Solar thermal energy utilizes the sun's heat to generate heat for boilers.

- Solar thermal collectors absorb solar radiation and transfer the heat to a fluid, which is then used as a heat source in boilers.

- Solar thermal energy can be used in combination with other energy sources to reduce the overall fuel consumption of boilers.

5. Geothermal Energy:

- Geothermal energy harnesses the heat from the Earth's crust to generate heat for boilers.

- Geothermal heat pumps or geothermal wells extract the heat from the ground and transfer it to a fluid that can be used for heating purposes.

- Geothermal energy is a sustainable and constant heat source, offering a reliable alternative to traditional fossil fuels.

6. Hydrogen:

- Hydrogen is a versatile energy carrier that can be produced from renewable sources through processes like electrolysis.

- Hydrogen can be used as a fuel in boilers, either by direct combustion or through fuel cells that generate heat and electricity.

- When produced using renewable energy sources, hydrogen has the potential to provide a carbon-free alternative for boiler operation.

7. Waste Heat Recovery:

- Waste heat recovery systems capture and utilize waste heat generated from industrial processes or exhaust gases from other equipment.

- The recovered heat can be used to preheat boiler feedwater, reducing the energy required for steam generation.

- Waste heat recovery is an energy-efficient solution that enhances the overall efficiency of boiler systems.

D. Waste Heat Utilization and Cogeneration

Waste heat utilization and cogeneration are strategies that maximize energy efficiency by capturing and utilizing the waste heat generated from industrial processes or power generation. These approaches help reduce energy waste, lower greenhouse gas emissions, and improve overall sustainability. Here's an overview of waste heat utilization and cogeneration:

1. Waste Heat Utilization:

- Waste heat is the heat generated as a byproduct of various industrial processes, such as exhaust gases from furnaces, flue gases from boilers, or hot process streams.
- Waste heat utilization involves capturing and utilizing this heat for productive

purposes instead of allowing it to dissipate into the environment.

- Common methods of waste heat utilization include:

- Heat Recovery Steam Generators (HRSG): Capturing waste heat to generate steam, which can be used for various applications, including electricity generation or process heating.

- Heat Exchangers: Transferring waste heat to other process streams or fluids to preheat feedwater, air, or other liquids.

- Organic Rankine Cycle (ORC): Utilizing waste heat to generate electricity through a thermodynamic cycle using an organic working fluid with a lower boiling point than water.

- Absorption Chillers: Utilizing waste heat for refrigeration or air conditioning purposes by driving an absorption cooling process.

- District Heating Systems: Directing waste heat to provide heating for neighboring buildings or industrial facilities.

2. Cogeneration (Combined Heat and Power, CHP):

- Cogeneration is the simultaneous production of useful heat and electricity from a single energy source.

- Cogeneration systems maximize energy efficiency by utilizing the waste heat generated during electricity generation.

- In conventional power generation, a significant amount of heat is wasted during the conversion of fuel to electricity. Cogeneration captures and utilizes this waste heat for various applications.

- Cogeneration systems can be implemented in different sectors, including industrial plants, commercial buildings, and district heating systems.

- The primary benefits of cogeneration include higher overall energy efficiency, reduced fuel consumption, lower emissions, and cost savings.

3. Benefits of Waste Heat Utilization and Cogeneration:

- Energy Efficiency: Waste heat utilization and cogeneration significantly improve energy efficiency by minimizing energy waste and utilizing available heat resources.

- Cost Savings: By utilizing waste heat for useful purposes, businesses can reduce their reliance on purchased energy, resulting in cost savings.

- Environmental Impact: Waste heat utilization and cogeneration reduce greenhouse gas emissions by maximizing the utilization of fuel and reducing the need for additional energy sources.

- Energy Security: Cogeneration systems provide a decentralized energy supply, enhancing energy security by reducing dependence on the grid.

- Resilience: Cogeneration systems can serve as backup power sources during grid outages, ensuring continuous operation and minimizing disruptions.

4. Considerations for Implementation:

- Feasibility Assessment: Conduct a thorough feasibility study to assess the technical, economic, and environmental viability of waste heat utilization or cogeneration systems.

- System Design: Optimize the design of the waste heat utilization or cogeneration system to match the specific requirements of the facility or application.

- Regulatory Compliance: Ensure compliance with local regulations and permits related to waste heat utilization, cogeneration, and power generation.

- Maintenance and Monitoring: Implement regular maintenance and monitoring programs to ensure efficient operation and performance of the waste heat utilization or cogeneration system.

- Integration with Energy Management: Integrate waste heat utilization or cogeneration systems with energy management systems to optimize overall energy efficiency and control.

Chapter 8: Future Trends In Boiler Technology

A. Advances In Boiler Design And Materials

Advances in boiler design and materials have significantly contributed to improved efficiency, reliability, and safety of boiler systems. Here are some notable advancements in boiler design and materials:

1. High-Efficiency Boilers:

 - Modern boiler designs focus on maximizing fuel efficiency and reducing energy waste.

 - Enhanced heat transfer surfaces, such as finned tubes or extended surfaces, increase the surface area available for heat exchange, improving boiler efficiency.

- Combustion optimization technologies, advanced burner designs, and intelligent control systems enable precise control of combustion parameters, resulting in more efficient and cleaner combustion.

2. Compact and Modular Designs:

- Compact and modular boiler designs allow for easier installation, maintenance, and expansion of boiler systems.

- Modular boilers consist of multiple smaller modules that can be combined to meet varying heat demands. This approach offers flexibility, scalability, and improved system reliability.

3. Flexible Fuel Capability:

- Advanced boiler designs accommodate a wider range of fuel options, including alternative and renewable fuels.

- Dual-fuel or multi-fuel boilers can seamlessly switch between different fuel sources, such as natural gas, oil, biomass, or hydrogen, optimizing fuel availability and cost efficiency.

4. Ultra-Low Emission Technologies:

- Advanced emission control technologies, such as selective catalytic reduction (SCR) and flue gas recirculation (FGR), reduce the emission of pollutants, including nitrogen oxides (NOx) and particulate matter.

- Low-NOx burners, staged combustion, and advanced air distribution systems help minimize emissions while maintaining efficient combustion.

5. Corrosion and Erosion Resistance:

- Improved materials and coatings enhance boiler durability and resistance to corrosion and erosion.

- High-temperature alloys, such as stainless steels, nickel-based alloys, and refractory metals, are employed in critical boiler components exposed to extreme heat and corrosive conditions.

- Protective coatings and advanced surface treatments further enhance resistance to corrosion, erosion, and fouling.

6. Water Treatment and Scale Prevention:

- Advancements in water treatment technologies and chemical additives help prevent scale formation, corrosion, and fouling in boiler systems.

- Improved water treatment practices, such as advanced filtration, reverse osmosis, and ion exchange, optimize water quality and minimize the potential for scaling and deposits.

7. Advanced Control and Monitoring Systems:

- Digital control systems and advanced sensors enable real-time monitoring and optimization of boiler performance.

- Integrated control systems provide precise control of combustion, temperature, pressure, and other parameters, optimizing boiler efficiency and ensuring safe operation.

- Remote monitoring and diagnostics enable proactive maintenance, early fault detection, and system optimization.

8. Insulation and Heat Recovery:

- Enhanced insulation materials and designs minimize heat loss and improve overall energy efficiency.

- Heat recovery systems, such as economizers, air preheaters, and condensing heat exchangers, capture waste heat from flue gases and utilize it for preheating combustion air or boiler feedwater.

9. Computational Fluid Dynamics (CFD) Modeling:

- CFD modeling and simulation techniques aid in the design and optimization of boiler systems.

- CFD analysis helps visualize and understand fluid flow patterns, heat transfer characteristics, and combustion behavior within the boiler, leading to improved performance and efficiency.

B. Digitalization And Automation In Boiler Operations

Digitalization and automation have revolutionized boiler operations by improving efficiency, safety, and control. Here are some key aspects of digitalization and automation in boiler operations:

1. Monitoring and Control Systems:

- Digital control systems provide real-time monitoring and control of various boiler parameters, including temperature, pressure, fuel flow, and emissions.

- Advanced sensors and instrumentation collect data and transmit it to the control system for analysis and decision-making.

- Automated control algorithms adjust combustion parameters, optimize fuel-air ratios, and maintain optimal operating conditions, resulting in improved efficiency and reduced emissions.

2. Remote Monitoring and Diagnostics:

- Boiler systems can be remotely monitored and managed using connected technologies and the Internet of Things (IoT).

- Remote monitoring allows operators and maintenance personnel to access real-time data, analyze performance, and identify potential issues.

- Predictive maintenance algorithms use historical and real-time data to predict equipment failures or performance degradation, enabling proactive maintenance and minimizing downtime.

3. Energy Management Systems (EMS):

- EMS platforms integrate data from multiple boilers and energy-consuming systems to optimize energy usage and reduce overall energy consumption.

- EMS solutions analyze data, identify energy-saving opportunities, and recommend operational adjustments to achieve energy efficiency goals.

- Energy dashboards and reporting tools provide insights into energy usage patterns, enabling operators to make informed decisions.

4. Integration with Building Management Systems (BMS):

- Boiler automation systems can integrate with BMS platforms to create a centralized control and monitoring solution for the entire facility.

- Integration with BMS allows coordinated control of HVAC systems, lighting, and other energy-consuming devices, optimizing overall energy efficiency.

5. Safety and Alarm Systems:

- Digitalization enables the implementation of advanced safety systems that monitor and detect abnormal operating conditions.

- Alarms and alerts notify operators in real-time about critical events, such as high-pressure situations, flame failure, or equipment malfunctions.

- Safety shutdown systems can automatically initiate emergency shutdown procedures in case of hazardous conditions, ensuring the safety of personnel and equipment.

6. Data Analytics and Optimization:

- Digitalization allows for the collection and analysis of large amounts of operational data to identify optimization opportunities.

- Data analytics tools can identify trends, patterns, and correlations to optimize combustion, minimize energy waste, and improve overall efficiency.

- Optimization algorithms can recommend operational adjustments, load distribution strategies, or fuel switching options to achieve maximum efficiency and cost savings.

C. Integration Of Renewable Energy With Boiler Systems

The integration of renewable energy with boiler systems offers several benefits, including reduced greenhouse gas emissions, increased energy efficiency, and enhanced sustainability. Here are some key aspects of integrating renewable energy with boiler systems:

1. Solar Thermal Systems:
- Solar thermal systems can be integrated with boiler systems to provide preheated water or steam, reducing the energy required for heating.
- Solar collectors capture solar energy and transfer it to a heat transfer fluid, which can then be used to supplement the boiler's heating capacity.

- Integration with solar thermal systems reduces the load on the boiler, resulting in energy savings and lower carbon emissions.

2. Biomass Combustion:

- Biomass, such as wood pellets, agricultural residues, or dedicated energy crops, can be used as a renewable fuel source in boilers.

- Biomass combustion systems can replace or supplement traditional fossil fuel burners, reducing the reliance on non-renewable resources.

- Co-firing biomass with conventional fuels allows for a gradual transition and provides flexibility in adjusting the fuel mix based on availability and cost.

3. Biogas Utilization:

- Biogas generated from anaerobic digestion of organic waste can be used as a fuel in boilers.

- Biogas combustion not only reduces emissions but also provides a means for waste management by utilizing organic waste streams.

- Proper treatment and purification of biogas are necessary to ensure its quality and compatibility with boiler systems.

4. Waste-to-Energy:

- Waste-to-energy technologies, such as incineration or gasification, convert waste materials into energy, which can be utilized in boilers.

- Combustion of waste materials reduces landfill waste, minimizes environmental pollution, and generates renewable heat or electricity.

- Advanced emission control systems are employed to ensure compliance with environmental regulations.

5. Heat Pumps:

- Heat pumps can be integrated with boiler systems to provide additional heating capacity using renewable energy sources.

- Air-source or ground-source heat pumps extract heat from the environment and transfer it to the boiler system for distribution.

- Heat pumps can reduce the energy consumption of the boiler by utilizing renewable heat sources, particularly in moderate climates.

6. Combined Heat and Power (CHP) Systems:

- CHP systems, also known as cogeneration, simultaneously produce heat and electricity from a single fuel source.

- By integrating renewable energy sources, such as solar or biomass, with CHP systems, the overall carbon footprint can be further reduced.

- The heat generated from the CHP system can be used for process heating or space heating, complementing the boiler's heat output.

7. Energy Storage:

- Integration of energy storage technologies, such as batteries or thermal storage systems, with boiler systems enhances the utilization of renewable energy.

- Excess energy from renewable sources can be stored and later used during periods of high demand or low renewable energy availability.

- Energy storage enables better management of renewable energy fluctuations and provides flexibility in optimizing boiler operations.

8. Control and Optimization:

- Advanced control systems and optimization algorithms can be employed to manage the integration of renewable energy sources with boiler systems.

- Intelligent control strategies can optimize the operation of boilers based on the availability of renewable energy, grid conditions, and demand profiles.

- Integration with energy management systems enables effective coordination between renewable energy sources, boiler operation, and energy storage.

D. Sustainability And Environmental Innovations

Sustainability and environmental innovations play a crucial role in the boiler industry by promoting energy efficiency, reducing emissions, and minimizing the environmental impact of boiler operations. Here are some key sustainability and environmental innovations in the boiler industry:

1. Low-Emission Burners:

  - Advanced low-emission burner designs, such as low-NOx burners, promote cleaner combustion by minimizing nitrogen oxide (NOx) emissions.

  - These burners use techniques like staged combustion, flue gas recirculation (FGR), and air-fuel ratio control to optimize combustion and reduce pollutant emissions.

2. Flue Gas Treatment Systems:

- Flue gas treatment systems, such as selective catalytic reduction (SCR) and electrostatic precipitators (ESP), help remove pollutants from the flue gas before it is released into the atmosphere.

- SCR systems use catalysts to convert nitrogen oxides (NOx) into nitrogen and water, significantly reducing NOx emissions.

- ESPs use an electric charge to capture and remove particulate matter (PM) from the flue gas, ensuring cleaner emissions.

3. Waste Heat Recovery:

- Waste heat recovery systems capture and utilize waste heat from boiler flue gases or other industrial processes, improving overall energy efficiency.

- Heat exchangers, economizers, and condensing heat recovery systems extract and utilize the waste heat for preheating combustion air or boiler feedwater, reducing energy consumption.

4. Carbon Capture and Storage (CCS):

- Carbon capture and storage technologies capture carbon dioxide (CO_2) emissions from boiler flue gases and other industrial sources.

- CO_2 is captured, compressed, and stored underground or used for industrial applications, preventing its release into the atmosphere and reducing greenhouse gas emissions.

5. Biomass and Bioenergy:

- The use of biomass as a renewable energy source in boilers reduces reliance on fossil fuels and contributes to carbon neutrality.

- Advanced biomass combustion technologies, such as fluidized bed boilers

and biomass gasification, enable efficient
and clean utilization of biomass fuels.

6. Energy Management and Optimization:

- Energy management systems and
optimization algorithms help monitor and
optimize boiler operations to maximize
energy efficiency and minimize waste.
- Real-time data analysis, predictive
modeling, and intelligent control strategies
enable proactive energy management, load
optimization, and demand-side management.

7. Renewable Energy Integration:

- Integrating renewable energy sources,
such as solar thermal, biomass, or biogas,
with boiler systems reduces the carbon
footprint and promotes a sustainable energy
mix.
- Combined heat and power (CHP)
systems and hybrid boiler systems enable

the efficient integration of renewable energy and contribute to decentralized and resilient energy systems.

8. Green Building Certifications:

- Boiler manufacturers and building projects can pursue green building certifications, such as LEED (Leadership in Energy and Environmental Design), to ensure sustainable and environmentally responsible practices.

- Green building standards promote energy-efficient designs, renewable energy utilization, and adherence to stringent environmental regulations.

9. Life Cycle Assessment (LCA):

- Life cycle assessment methodologies evaluate the environmental impact of boilers and associated processes throughout their entire life cycle, from raw material extraction to disposal.

- LCA helps identify areas for improvement, guide sustainable product design, and inform decision-making by considering environmental factors.

IX. Glossary Of Boiler Terminology

Here's a glossary of common boiler terminology:

1. Boiler: A closed vessel in which water or other fluid is heated to generate steam or hot water.

2. Combustion: The process of burning fuel in the presence of oxygen to release heat energy.

3. Fuel: Any material that can be burned to produce heat energy, such as coal, oil, natural gas, biomass, or solid waste.

4. Heat Exchanger: A device that transfers heat from one fluid to another without direct contact between the two fluids.

5. Steam: Water vapor generated by boiling water in a boiler. It finds frequent application in heating, power generation, and industrial operations.

6. Hot Water: Water heated by a boiler to a specified temperature and used for heating or domestic purposes.

7. Pressure: The force exerted per unit area, typically measured in pounds per square inch (psi) or bars.

8. Temperature: The degree of hotness or coldness of a substance, usually measured in degrees Celsius (°C) or Fahrenheit (°F).

9. BTU (British Thermal Unit): A unit of measurement for heat energy. It is the heat capacity needed to elevate the temperature

of one pound of water by one degree
Fahrenheit..

10. Efficiency: The ratio of useful output
energy or work to the input energy or fuel
consumed. Boiler efficiency is the measure
of how effectively the boiler converts fuel
into usable heat.

11. Blowdown: The process of removing a
portion of the boiler water to control the
concentration of impurities and maintain
water quality.

12. Soot: A black, powdery substance
composed of carbon particles that result
from incomplete combustion of fuel.

13. Scale: Deposits of minerals, such as
calcium or magnesium, that accumulate on

the inner surfaces of boiler components due to water impurities.

14. Safety Valve: A device designed to release excess pressure from the boiler to prevent overpressure and ensure the safety of the system.

15. Flame Failure: The condition when the flame in a burner or combustion chamber is extinguished or fails to ignite.

16. Water Level: The height of the water in the boiler measured using water level gauges or electronic sensors.

17. Condensate: The liquid formed when steam cools and changes back into water after releasing its heat energy.

18. Draft: The flow of air or flue gases in the boiler, which is essential for proper combustion and removal of combustion products.

19. Deaerator: A device used to remove dissolved gases, such as oxygen and carbon dioxide, from the boiler feedwater to prevent corrosion and improve efficiency.

20. Flame Scanner: A sensor that detects the presence and stability of the burner flame and provides feedback to the control system for flame monitoring and safety.

21. Economizer: A heat exchanger that preheats the boiler feedwater using waste heat from the flue gases, improving overall efficiency.

22. Superheater: A heat exchanger that increases the temperature of the steam beyond its saturation point, producing superheated steam with additional heat energy.

23. Turndown Ratio: The ratio of the maximum to minimum heat output capacity of a boiler. It represents the boiler's ability to modulate its output to match the load demand.

24. Water Treatment: The process of treating boiler water to remove impurities, prevent scale formation, and maintain proper water chemistry.

25. Corrosion: The deterioration of boiler components due to chemical reactions between the metal surfaces and the surrounding environment, typically caused by water or combustion byproducts.

26. Flue Gas: The exhaust gases produced by the combustion process and discharged through the boiler's flue or chimney.

27. Low-Water Cutoff: A safety device that shuts down the boiler if the water level drops below a specified minimum.